MW00509456

ROSEN ✔ *Verified*
NATURAL DISASTERS

FLOODS

Benjamin Proudfit

ROSEN
PUBLISHING

Published in 2023 by The Rosen Publishing Group, Inc.
2544 Clinton Street, Buffalo, NY 14224

First Edition

Editor: Kristen Nelson
Book Design: Michael Flynn

Library of Congress Cataloging-in-Publication Data

Names: Proudfit, Benjamin, author.
Title: Floods / Benjamin Proudfit.
Description: Buffalo, New York : Rosen Publishing, [2023] | Series: Rosen
  verified: natural disasters | Includes index.
Identifiers: LCCN 2022023374 (print) | LCCN 2022023375 (ebook) | ISBN
  9781499469523 (library binding) | ISBN 9781499469516 (paperback) | ISBN
  9781499469530 (ebook)
Subjects: LCSH: Floods--Juvenile literature. | Flood control--Juvenile
  literature.
Classification: LCC GB1399 .P76 2023 (print) | LCC GB1399 (ebook) | DDC
  551.48/9--dc23/eng/20220524
LC record available at https://lccn.loc.gov/2022023374
LC ebook record available at https://lccn.loc.gov/2022023375

Manufactured in the United States of America

Some of the images in this book illustrate individuals who are models. The depictions do not imply actual situations or events.

CPSIA Compliance Information: Batch #CWRYA23. For further information contact Rosen Publishing at 1-800-237-9932.

Find us on

CONTENTS

FLOODS HAPPEN EVERYWHERE

Imagine a heavy rainstorm that goes on for days. Depending on where you live, this can have different effects. However, if the rain is heavy enough and goes on for long enough, a flood can happen—no matter where you live. In some areas, the land may be able to absorb, or take in, a lot of rain. In cities, though, less rain may cause a flood sooner!

A flood is when water rises and overflows onto land that is usually dry. This can happen near a body of water or an area where water doesn't normally pool. A flood becomes a natural disaster when it causes great harm. The harm may be **damage** to homes, businesses, or farms. It also may include loss of life.

A flood can happen almost anywhere in the world! In the United States, flooding may occur in all 50 states.

A flood can be a natural disaster at any depth. It may be just inches deep and fast moving. A flood can be so deep it rises to the roof of a house! A flood that lasts only a few hours can be a disaster. Floods can last for days, weeks, or months too. No matter where or when a flood happens, it can be costly to the community it hits.

BUILDING BY WATER

Early **civilizations** were built near bodies of water. People used the water for washing and drinking. They watered their crops and used it to travel. Floods have always come with living near water. Sometimes these floods were good for the people living there. The yearly overflow of the Nile River in Egypt helped make the land fertile, or good for growing crops.

NILE RIVER

But deadly floods are always a risk. Millions of people died in 1931 because the Yangtze River in China flooded. Three floods of China's Yellow River in 1887, 1931, and 1938 killed millions of people too.

GREAT FLOODS

Many **cultures** have a story about a flood so great that it covered Earth! The most familiar of these might be the story of Noah and the Ark. This story is in the Bible. In it, Noah and his family gathered two of each animal on Earth and rode in an ark, or large boat, until the flood ended.

CITIES ON THE MISSISSIPPI RIVER

Major cities in the United States were built on the Mississippi River. They are at risk of flooding if the Mississippi overflows!

TYPES OF FLOODS

There are several types of floods. In a river flood, the water in a river rises higher than its banks and overflows. Coastal flooding occurs when water from the ocean comes onto land higher than normal. These floods can be shallow, or not deep. A storm surge is a sudden rise in water on the coast caused by a storm like a **hurricane**.

Inland flooding occurs away from the coast. It may be caused by a lot of rain or a river overflowing. Flash floods are floods that occur after just a few hours of heavy rainfall or another sudden increase in water.

FAST FACT

Storm surges can rise 20 feet (6 m)! They often cause great damage.

Cities that have a lower **elevation** have more flooding. In the United States, Charleston, South Carolina, and Savannah, Georgia, have some coastal flooding every year. In 2015, Charleston had 38 occurrences of "sunny-day" flooding, or coastal flooding having to do with higher-than-average tides. Tides are the upward and downward movement of the ocean.

In Charleston, coastal flooding has become a common part of the weather.

WHAT CAUSES FLOODS?

Many floods are caused by heavy rainfall. There may be very rainy weather over a long period of time. The ground on which the rain falls is unable to absorb any more water, or it fills a body of water to the point of overflowing.

Heavy rainfall is more likely to cause a flood in places with many paved surfaces, like cities. The rainwater runs off of sidewalks and streets. They cannot absorb it. Having **steep** hills or mountains nearby increases flood risk to people too. The rain flows downhill toward homes and businesses.

MONSOON SEASON

South and Southeast Asia have a lot of rain during the wet monsoon season. This is from about May to September each year. There is also a dry monsoon season. Monsoons are winds that blow in a certain direction and affect the weather. In 2019, the flooding caused by the wet monsoon season in Bangladesh, India, Myanmar, and Nepal killed at least 600 people.

Bangladesh is a county to the east of India. It is one place where increasingly heavy rains during monsoon season have caused bad flooding.

Some of the worst flooding is caused by other natural disasters. Hurricanes cause incredible amounts of rain, which leads to flooding. The high winds often cause storm surges, which flood coasts.

Tsunamis are huge waves in the ocean that are commonly caused by underwater **earthquakes** and other movements of the ocean floor. When a tsunami reaches land, waves up to 30 feet (9.1 m) high can crash on the coast. They cause many miles of flooding and lots of damage.

HURRICANE FLOODING

THESE HURRICANES CAUSED SOME OF THE WORST FLOODING THE UNITED STATES HAS EVER SEEN.

HURRICANE NAME	YEAR
Labor Day Hurricane	1935
Hurricane Camille	1969
Hurricane Agnes	1972
Hurricane Katrina	2005
Hurricane Harvey	2017

A huge tsunami hit Indonesia in 2004. The flooding caused terrible damage, including **contaminating** drinking water for a long time after.

FLASH FLOODS

Flash floods are caused by many of the same events as other floods. They just happen more quickly! In order to be called a flash flood, the flooding has to start within six hours or less of the rainfall or other cause. Big storms like hurricanes cause flash floods, but they have many other causes too.

Flash floods may happen when it's not raining. A **dam** can break and release, or let go, too much water, causing a flash flood. This happened in Johnstown, Pennsylvania, in 1889. More than 2,000 people died.

FAST FACT

Johnstown had two more huge floods, one in 1936 and one in 1977.

14

BARE LAND

Land without much vegetation, or plant life, is more likely to flood. **Deforestation** and wildfires leave bare land where a flash flood could **develop** if it rains enough. The plants and soil aren't there to absorb the water moving across the land. The water may also pick up **debris** and cause damage that way.

It may take a week's worth of rain for a flood to occur. A flash flood develops in hours.

Melting snow and ice jams can also cause flash floods. As weather warms, snow starts to melt. It adds to the water in streams and rivers. This melting snow can break up ice that's built up over the winter. Ice moves downstream and can get caught—or jammed—against bridges. Water builds up behind the ice jam. When the ice comes loose or melts, the water rushes out, sometimes causing a flash flood.

One other way a flash flood may occur has to do with **storm drains**. Debris washed into them may plug them up. Too much rain makes them overflow, just like any waterway. Places that are low, like basements, can flood fast when this happens.

THE BREAKUP

It's hard to guess when an ice jam will break up. Then, it may cause flooding, or it may not. In general, a few warm, rainy days in a row make the conditions right for a flash flood to occur from an ice jam breaking up.

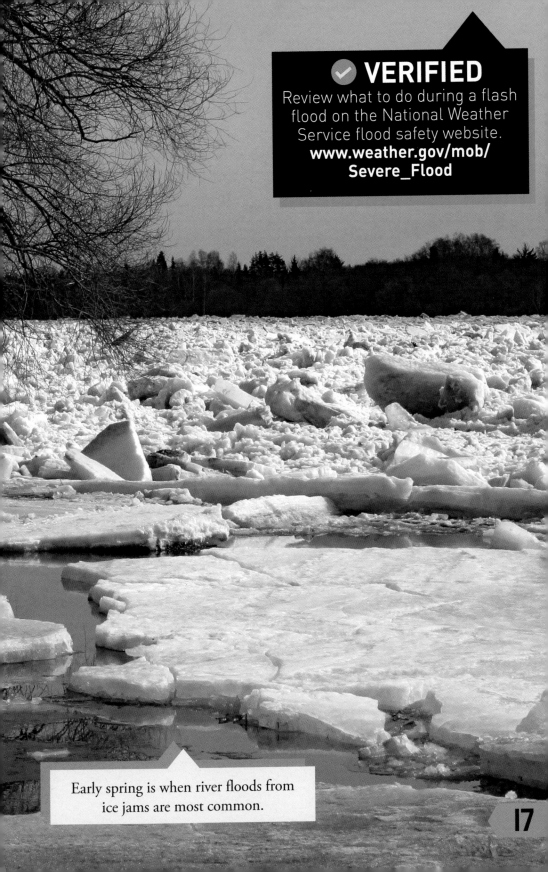

Early spring is when river floods from ice jams are most common.

PREDICTING FLOODS

Some natural disasters cause so much harm and so many deaths because they happen without warning. Meteorologists and hydrologists watch for places where floods may occur. Sometimes, they figure it out. But, flash floods may happen without **detection**.

Some questions meteorologists ask when predicting flooding include:

- **How much rain is falling?**
- **How fast is the storm moving?**
- **How long does the storm stay over one area?**
- **How much rain has the area had lately?**
- **What is the land like where the storm is?**

SCIENTISTS AT WORK

Meteorologists are scientists that study weather. They predict, or use **data** and knowledge to guess, what the weather will be. This includes predicting when storms may hit. They try to figure out how big storms will be and where they will move.

Hydrologists are scientists who study how water moves around Earth and its atmosphere, or the gases surrounding Earth. They keep track of water levels in rivers and lakes. They study how more precipitation might change those levels. Precipitation is any water that falls to the ground, such as rain or snow.

Meteorologists study past floods and what kinds of weather caused them to develop. They can use this knowledge to better predict future floods.

Scientists use many tools when predicting weather that could lead to a flood. Radar is a tool that uses radio waves to find out where something is and how fast it's moving. It's used to measure how fast and in which direction precipitation is falling. The measurements of many radar systems are used together to get a full picture of the weather in a place.

Satellites are used to watch areas that don't have many weather predicting tools nearby. They also help scientists see weather over a big area, such as across oceans and mountains.

FAST FACT
Rain gauges are tools that measure rainfall in one place.

Oklahoma City

On weather radar maps, rain is shown in green. Rain that's a bit heavier is in yellow. Very heavy rain is in red.

DOPPLER RADAR

Doppler radar is the tool used to follow weather systems. Doppler can show meteorologists what kind of precipitation is falling. It can show where it is and how it moves. This can be used to figure out how that weather might move or change.

WATCH OR WARNING?

Meteorologists may be able to predict floods days or weeks ahead of time. They may know where the conditions are right for a flash flood to happen. But, they can't be totally sure if or when it will happen.

The National Weather Service tells the people living near an area at risk for a flood when it can. A flood advisory is issued when flooding may happen, but it won't be a problem for most people. A flood watch means the conditions are right for flooding to happen. A flood warning means flooding is about to happen or is already happening.

SEE AND HEAR IT

You may hear about a flood advisory, watch, or warning in a few ways. You may see it scroll on the bottom of your TV screen if you are watching TV. You may hear it on the radio. Finally, some smartphones have an alert system that goes off when a flood or flash flood warning is issued for the area.

●●○○○ Verizon LTE   🔋 ⏰ 🌙 🔆 58% 🔋

1:30
Thursday, February 14

⚠ EMERGENCY ALER...   1m ago

Emergency Alert
Flash Flood Warning this area til 3:45
PM PST. Avoid flood areas. -NWS
Slide to open

Press home to open

✅ VERIFIED
Get ready for emergency situations using the tips and tools on this website.
**www.ready.gov/kids/
be-ready-kids**

FIND OUT IF YOUR HOME IS AT HIGH RISK OF FLOODING.

MAKE AN EMERGENCY KIT WITH WATER, FOOD, A FIRST AID KIT, FLASHLIGHTS, AND OTHER IMPORTANT ITEMS.

MAKE A PLAN FOR FLOODING

CHOOSE A SAFE PLACE FOR YOUR FAMILY TO MEET IF YOU AREN'T TOGETHER.

LISTEN FOR FLOOD WATCHES AND WARNINGS AND FOLLOW ANY DIRECTIONS GIVEN ABOUT THEM.

Take steps before a flood happens to be ready.

FLOOD PREVENTION

Predicting where and when floods may happen has gotten better as weather **technology** has improved, or gotten better. Flood **prevention** has also improved.

Dams are structures built to hold water back for people's use. This includes flood control. Dams run across a body of water. If too much water builds up behind a dam, water can be let out of the dam in a controlled way.

Levees are walls built up along waterways. They are created to stop water from overflowing a waterway's banks. Levees are often made of earth or other natural matter. Sometimes, they are made stronger by adding plastic, metal, or concrete.

Engineers are people who build and plan systems and structures. They are the ones who find new and better ways to build dams and levees. After a flood caused by the failure of dams or levees, they try to figure out what went wrong. They will try to fix it in the future.

Dams have been used for flood control since the 1940s.

LEVEE AND LOCKS AT LAKE PONTCHARTRAIN

New Orleans, Louisiana, is built on Lake Pontchartrain near the Gulf of Mexico. It faces hurricanes that could cause flooding every year. It is known for its levee system—and the system's huge failure in 2005. That year, Hurricane Katrina caused a storm surge and rainfall that put much of the city underwater. The levees failed in more than 50 places during that storm.

The U.S. government put billions of dollars into updating the levees around New Orleans. In 2021, they faced a strong hurricane in Hurricane Ida. The city saw no flooding. The levees held!

HURRICANE KATRINA'S FLOODING

| ABOUT 80 PERCENT OF NEW ORLEANS WAS FLOODED. | STORM SURGES REACHED 30 FEET (9.1 M). | FLOODWATERS WERE UP TO 15 FEET (4.5 M) DEEP. |

IN THE NETHERLANDS

About one-third of the Netherlands is below sea level. For thousands of years, those who live in the Netherlands have built dikes. These built-up walls along the water keep it out. In the 1950s, dikes didn't stop major flooding. Since then, the country has improved their dikes and now also has levees, dams, and other flood-prevention measures in place

It rains about 200 days a year in the Netherlands!

SAFE AT HOME?

Levees and dams can often keep floods from storm surges or overflowing rivers from harming an area. But floods still can happen.

Governments put guidelines in place for buildings to be safe during flooding. There may be even more guidelines on **floodplains**. One of these is to make sure buildings are built up high enough, even if they are on land with low elevation. Certain building materials are considered more flood safe than others too.

PROTECT YOUR PROPERTY

People who own a home have homeowner's insurance. They pay money to an insurance company each year. In turn, if something bad happens to their home, like a fire, the insurance company will help pay for repairs and damages. But most home insurance plans don't include flood insurance. People living in places at high risk of flooding may have to pay more money to add this to their plan.

The World Health Organization reports that 2 billion people were affected by flooding around the world between 1998 and 2017.

It's well known where floods are likely to happen. Communities built along rivers or the coast are likely to have some flooding. Places in deep valleys are at risk for flooding. And some places have had big floods many times. People living there may have been through more than one flood and lost more than one home! So why do they still live there?

People may choose to stay in a place because their family lives there. They may feel connected to their community and not want to start a new life somewhere else. They may choose to rebuild a home on a floodplain to keep these connections.

MOVING FROM A FLOODPLAIN

In some places, the U.S. government is trying to move people away from floodplains. It would save billions of dollars in damage and natural disaster **relief**. Many plans have the government buying homes on a floodplain. They have those living there move to new homes. Then, they knock down the old homes and let nature take over. This action doesn't prevent floods, but it can reduce the harm done to people and homes over time.

Whole neighborhoods have been knocked down and the people moved after floods. Their homes are rebuilt on higher ground.

DURING A FLOOD

When a flood warning is issued, people may be asked to evacuate. This means they have to leave their homes, often very quickly. There isn't always time for evacuation. However, there are still ways to stay safe.

Stay away from low-laying areas during a flood. People who are camping or live in a place at a low elevation should get to higher ground. No one should walk or drive through floodwaters. The water can be deeper than it looks. It can move very quickly. There may be electrical wires hidden in the water.

HIGHER GROUND

What is higher ground? It's any place that is higher than the area around it. If a person is outside when a flood hits, the top of a hill can work. Going to a higher floor in a building or your home is also moving to higher ground.

People who stay home because they can't or don't want to evacuate may risk their lives during a flood.

HISTORIC FLOODS

The Mississippi River Flood of 1927:
This flood occurred after months of heavy
rainfall. The Mississippi overflowed and levees
along it broke down. The flood eventually
covered 23,000 square miles (60,000 sq km)
with water! About 250 people died, making it
one of the deadliest floods in U.S. history.

The Buffalo Creek Dam Flood: In
1972, Southern West Virginia had one of the
worst floods ever in the United States. Three
dams had been built to hold back wastewater
from coal mines. They had been poorly built.
A rainstorm hit, and the water was too much
for the dams. Each failed. Almost 130 million
gallons (490 million l) of water and other
matter flooded out. The flood killed 125
people and tore down about 550 homes.

Canyon Lake Flood: Canyon Lake is a reservoir behind Canyon Dam in central Texas. When it rained heavily for many days in the dry summer of 2002, the reservoir overflowed. The floodwaters were so strong and fast that they formed a canyon in just three days!

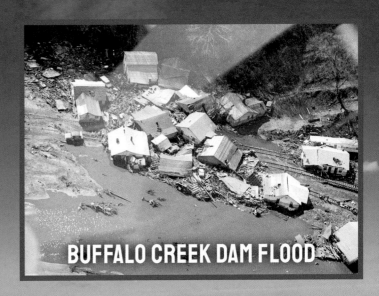

BUFFALO CREEK DAM FLOOD

CANYON DAM

A reservoir is a human-made lake that can hold a large amount of water.

AFTER A FLOOD DISASTER

When a flood large enough to be called a natural disaster occurs, help is needed right away. Groups like FEMA and the Red Cross often organize relief efforts. The National Guard may be called in to help too.

First, people who have been stranded by the floodwaters need to be rescued. Helicopters, boats, and rafts are used to find and help these people get to safety. Medical care is given to anyone who is hurt. Places are set up for people whose homes have been flooded, so they have somewhere to go.

FAST FACT

In the United States, there are more than 100 deaths each year caused by flooding.

Floods often cause **sewage** and other harmful matter to contaminate an area's water supply. Clean drinking water is brought to areas where a flood disaster has occurred.

✓ **VERIFIED**
Find out more about how the Red Cross helps after a flood disaster here.
www.redcross.org/about-us/ our-work/disaster-relief/ flood-relief.htm

In 2021, terrible flooding in Tennessee meant relief was badly needed.

Communities have a long cleanup after flooding. They clean up debris, sewage, and other harmful matter that has been caused or brought in by the flood. Downed power lines and other safety concerns have to be addressed before anyone who lives there can return.

Rebuilding a city or neighborhood damaged by flooding is a huge job. Not only do buildings need to be repaired, but transportation like buses and cars may need to be replaced. Crops may need replanting.

DISPLACED PEOPLE

In 2018, more than 5 million people around the world were displaced, or forced to leave their homes, because of flooding. Displaced people are one of the major long-term effects of flooding. They need places to stay and basic supplies like food and clothing. Many are unsure of what to do next.

In April 2022, very heavy rains in South Africa led to a flood disaster. The cleanup and rebuilding will take years to complete.

HEALTH CONCERNS

Health problems are another long-term effect of floods. Floodwater can cause diseases, or illnesses, to spread quickly. The flood may keep health care workers like doctors from being able to get to people who are sick or hurt. This can lead to bigger health problems and even death.

In addition, everyone who survives a disaster may have trouble with their mental health. It can make a person feel anxious, or nervous. They may feel upset and sad a lot. These feelings can last a long time.

FAST FACT

About 75 percent of flood disaster deaths are drownings, or deaths caused by being underwater too long and unable to breathe.

It can be hard to get needed health care following a natural disaster like a flood.

FLOODS AND CLIMATE CHANGE

Climate change means more flood risk in more places. Rising ocean levels mean coastal flooding will happen more often. Stronger hurricanes will drop more rain, causing greater flooding.

People around the world are trying to slow climate change. They are trying to increase the use of power that doesn't harm Earth. They encourage people to recycle and reuse things they already have instead of throwing everything away. Governments are also working together toward these goals. But some scientists say it's too little, too late.

IT'S HERE

South Africa got almost as much rain as it gets in a year in just two days in April 2022. It caused incredible flooding that killed hundreds of people. Meteorologists didn't predict this weather. The country's president, Cyril Ramaphosa, said: "It is telling us that climate change is serious, it is here."

WHAT IS CLIMATE CHANGE?

Climate change is the long-lasting change to Earth's weather. Part of this is global warming, or the slow warming of Earth due to many human activities. Scientists believe it is causing storms to be bigger and cause more damage.

Many people speak out about climate change. They don't want disasters like floods to become worse and more common.

KNOW THE FACTS

Floods can happen anywhere if the conditions are right. And because of climate change, floods of great size may happen more often. People who live near water, especially on a floodplain, need to be the most prepared. In some cases, that may mean moving to a neighborhood on higher ground. In the future, the U.S. government may move thousands of people out of floodplains to avoid disaster.

All of this doesn't mean people should be scared of flooding all the time. Meteorologists are hard at work right now tracking storms. They are taking data to predict where the next big storm will hit. They will issue advisories, watches, and warnings as soon as they can if they believe flooding is going to occur. Engineers, too, are working to make sure flood prevention measures are up-to-date and safe. They work to find new ways to keep flooding from harming communities.

DISASTER PREPAREDNESS
CHECKLIST

☐ FIRST AID KIT
☐ FLASHLIGHT, RADIO AND SPARE BATTERIES
☐ BLANKETS, CLOTHES

Floods are scary natural disasters. Make sure your family is ready!

Sometimes a flood cannot be stopped. Having a family emergency plan for natural disasters like floods is the first step to being prepared. Understanding this natural disaster, and what to do should one occur, is the next.

GLOSSARY

civilization: Organized society with written records and laws.

contaminate: To pollute something.

culture: The beliefs and ways of life of a group of people.

dam: A structure built across a river or stream to control the flow of water.

damage: Harm; also, to cause harm.

data: Facts and figures.

debris: The remains of something that has been broken.

deforestation: The cutting down of all trees in an area.

detection: The act of finding something out or noticing something.

develop: To form, grow, and change.

earthquake: A shaking of the ground caused by the movement of Earth's crust.

elevation: Height above sea level.

floodplain: Flat land next to a river that often floods.

hurricane: A powerful storm that forms over water and causes heavy rainfall and high winds. Typhoons and cyclones are other names for the same kind of storm, depending on where they occur and what direction the winds are moving.

prevention: The act of stopping something from happening.

relief: Aid that is given to people following a natural disaster.

satellite: An object that circles Earth in order to collect and send information or aid in communication.

sewage: Waste matter from buildings that is carried away through sewers.

steep: Almost straight up and down.

storm drain: A drain that carries water away from the street or other place.

technology: The use of science, engineering, and other industries to invent useful tools or to solve problems; also, a machine, piece of equipment, or method created by technology.

INDEX